IMAGES
of America

THE SEABEES AT GULFPORT

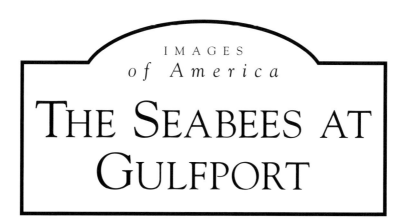

IMAGES
of America

THE SEABEES AT GULFPORT

Gina L. Nichols
Foreword by Capt. William Hilderbrand

ARCADIA
PUBLISHING

Published by Arcadia Publishing
Charleston SC, Chicago IL, Portsmouth NH, San Francisco CA

Printed in the United States of America

Library of Congress Catalog Card Number: 2007933994

For all general information contact Arcadia Publishing at:
Telephone 843-853-2070
Fax 843-853-0044
E-mail sales@arcadiapublishing.com
For customer service and orders:
Toll-Free 1-888-313-2665

Visit us on the Internet at www.arcadiapublishing.com

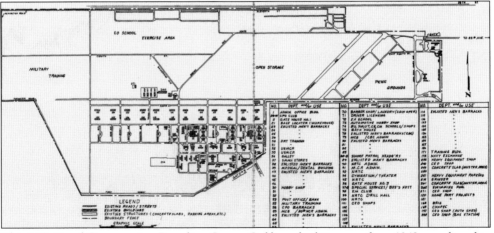

This U.S. Naval Construction Battalion Center Gulfport facilities map from 1942 was altered to illustrate the main buildings in 1966.

CONTENTS

ACKNOWLEDGMENTS

The author wishes to thank her friends, family members, and son, Hunter, who supported her during this venture. Special thanks go to Rear Adm. Michael Johnson, Vice Adm. Michael Loose, and John Hughes for their support of the Seabee museum archive. Without their support of the archival project, this book would not have been possible. Last, but certainly not least, thanks to the U.S. Navy Seabee Museum staff, volunteers, and patrons for their unending love and support of the Seabee history, which made this book worth creating.

All of the photographs in this book are from the U.S. Navy Seabee Museum and the Naval Media Center's Navy Newsstand. The U.S. Navy Seabee Museum archive collection, located at Construction Battalion Center Port Hueneme since its inception in 1961, is the official repository for the U.S. Naval Construction Force records. Official U.S. Navy, Marine Corps, and Air Force photographers captured all photographs presented here, and all fall under public domain.

The views, opinions, and statements presented in this book are those of the author and do not necessarily represent nor are they endorsed by the U.S. government, the Department of Defense, or the U.S. Navy.

FOREWORD

The story of the Seabee base at Gulfport, Mississippi, is one of crucial support to the navy's Seabees. These Seabees go to some of the most remote places on earth, perform incredible tasks under unbelievable conditions, and never ask why or to be thanked. Their only question is, "What else needs to be done?"

With the creation of the Seabees in early 1942 came the need for bases from which to ship the gargantuan amounts of construction materials they would need to build the overseas bases necessary to fight World War II. Gulfport was the location selected for one of three such bases created for this purpose, and enormous quantities of supplies were shipped to the Caribbean, Central and South America, and other places.

Dormant after World War II, Gulfport served in the late 1940s, 1950s, and early 1960s primarily as a site for warehousing construction equipment and huge amounts of war reserve materials. In 1966, it sprang into full operation again and welcomed Seabee battalions newly commissioned to serve in Vietnam. Expanding its responsibilities of World War II, it added the role of a home port to its mission. Today it is home to over half of the navy's active Seabee battalions and regiments and continues to relentlessly focus on its mission to be the best at supporting the Seabees as they work around the world, providing construction support for all the armed services and humanitarian assistance wherever required. The actions of the base and its Seabees in responding to the havoc wreaked by Hurricane Camille in 1969 and Hurricane Katrina in 2005 have earned both the base and the Seabees a special place in the hearts of Gulf Coast residents.

A special thank-you goes to Gina Nichols for her tireless effort in researching and helping tell the story of the Gulfport Seabee base and the countless tens of thousands of Seabees who have passed through its gates.

—William C. Hilderbrand
CAPT, CEC, USN (Ret.)
President
CEC/Seabee Historical Foundation

INTRODUCTION

The U.S. Naval Construction Forces, best known as the Seabees, were established in 1942 to meet the wartime need for trained uniformed men to perform construction work in combat zones. The Seabees performed an essential part in winning one of the greatest wars in history.

In 1939, the navy's Bureau of Yards and Docks began to prepare for the construction of advance bases in remote areas, a necessary part of waging a global war. Realizing that shipping enormous quantities of supplies and equipment abroad would overburden existing port facilities, officials decided in early 1942 to construct an advance base depot on each coast and a third on the Gulf Coast to serve as outlets for materials required overseas.

Gulfport became the logical position for the third base since it was located at the intersection of two major railroads, had a large municipal pier for shipping, and had large areas for storage, as well as being near the Caribbean sites of several proposed advance bases. On June 2, 1942, Advance Base Depot Gulfport was established and the first Seabees started coming through the gates. In October 1942, Advance Base Receiving Barracks Camp Hollyday was established at Advance Base Depot Gulfport. Camp Hollyday served as the personnel receiving and processing activity for deployed Naval Construction Force units and was responsible for training new personnel in transit.

During World War II, the Seabees built the navy's bases around the world and paved the roads to victory in the Atlantic, Alaskan, and Pacific theaters with their limitless construction skills. Their accomplishments during the war are legendary, including building over 400 advance bases, 111 major airfields, 441 piers, 2,558 ordnance magazines, hospitals to serve 70,000 patients, and housing for 1.5 million men. Nearly 325,000 men, master artisans and the most proficient of the nation's skilled workers, paved the road to victory for the Allies. They served on six continents and more than 300 islands, suffered more than 300 combat deaths, and earned more than 2,000 Purple Hearts.

In late 1945, as World War II activities ended, the Advance Base Depot was re-designated as a U.S. Naval Storehouse. In May 1952, the base was renamed Naval Construction Battalion Center (CBC) Gulfport under the administration of the Bureau of Yards and Docks. Two other component activities were simultaneously brought into being: the U.S. Naval Advance Base Storage Depot and U.S. Naval Construction Equipment Depot, which provided storage and maintenance of construction equipment. For 14 years, CBC Gulfport remained a Construction Battalion Center without any construction battalions.

In March 1966, the navy's escalating commitment for construction forces in Southeast Asia led the way to an increased mission for the center. Between 1965 and 1968, the total number of Naval Mobile Construction Battalions rose from 10 to 21, and nearly every battalion had grown from half strength to a full strength of 800 men. Naval Mobile Construction Battalions 62, 74, 121, 128, and 133 were commissioned at Gulfport from 1966 to 1967 to aid in meeting construction demands in Vietnam and Thailand. In 1969, Reserve Naval Mobile Construction Battalions 12 and 22 were activated in Gulfport to serve in Vietnam. The staff for the Naval Construction

Battalion Center expanded simultaneously to 183 military and 523 civilian personnel to support approximately 4,200 Seabees.

After Hurricane Camille cut an angry path across the Gulf Coast in August 1969, the Seabees were the first to go into action. Naval Mobile Construction Battalions 121 and 128 and the Naval Construction Training Center performed search-and-rescue operations, evacuations, power and communication restoration, and road clearance.

As the Seabees entered the post-Vietnam era, they found themselves employed in major peacetime projects that had been deferred or neglected because of wartime priorities. The post-Vietnam Seabees were involved in new construction frontiers in Diego Garcia, the Trust Territory of the Pacific Islands, Antarctica, and on the ocean floor itself. Civic Action Teams in the Trust Territory served on the islands of Ponape, Truk, Palau, Kusaie, and Yap building roads, dispensaries, water tanks, bridges, and public buildings.

On August 17, 1990, ten days after the commencement of Operation Desert Shield, the first Seabees arrived in Saudi Arabia. By October 18, 1990, Seabee mobilization was complete and encompassed 2,410 Seabees, 1,131 pieces of equipment, and 12,000 short tons of materials. Seven active battalions, two construction battalion units, and three full reserve battalions and portions of two others were reactivated in support of the Persian Gulf War.

In the summer of 1992, Seabees were called on to provide recovery assistance to the hurricane-devastated area of Homestead, Florida, following Hurricane Andrew. Seabees were also vital to the humanitarian efforts in Somalia during Operation Restore Hope in 1992–1993. In 1994, the Seabees provided assistance to the Haitian relief effort at Guantanamo Bay, Cuba. On Christmas Day 1995, Seabees arrived in Croatia to support the army by building camps as part of Operation Joint Endeavor, the peacekeeping effort in Bosnia-Herzegovina. In response to Hurricane Ivan's destruction at Naval Air Station Pensacola, Seabees deployed to Florida in September 2004 and were greeted with cheers as they arrived with heavy equipment and chain saws to clear away hurricane debris, repair roads, erect tents, and help their fellow service members.

More recently, the Seabees have been instrumental in providing support during Operation Enduring Freedom in Afghanistan and Operation Iraqi Freedom by repairing tarmac damage, cutting steel plating to enhance tactical vehicle armor, patrolling the streets of Fallujah prior to Iraq's historic democratic elections, building schools and hospitals, and implementing improvements to local village water, electricity, and sanitation facilities.

The Seabee motto *Construimus Batuimus* ("We Build, We Fight") and their concise expression "Can Do," which is their stock reply when asked if they can perform any job, summarizes their attitude and approach to any construction project. Navy Seabees deploy around the world to provide construction support for U.S. forces as well as humanitarian assistance and disaster recovery. Using Seabee "Can-do," they have been directly involved in projects that have assisted millions of people from the Thailand tsunami victims to the refugees in camps in Somalia, Bosnia, Vietnam, Kuwait, South America, and Kurdistan. The Seabees at Gulfport also support disaster recovery efforts closer to home. Following Hurricane Katrina, the Seabees at Gulfport were among the first to respond. While their own homes were decimated and their families were living in warehouses, the Seabees aided local residents, cleared streets, rescued pets, and restored Gulfport's infrastructure.

One

1942–1945
ADVANCE BASE DEPOT GULFPORT

Sailors on liberty from Naval Training Center leave through the Broad Avenue Gate (Gate No. 3), pictured here in 1944. Sailors stationed at Gulfport visited the Gulf Coast cities while on liberty and leave. Liberty buses traveled periodically to and from New Orleans, Biloxi, and Long Beach, allowing the men a chance to visit each city.

An aerial view of Advance Base Depot and Naval Training Center, Gulfport, is pictured here in 1944. On November 13, 1942, the Advance Base Receiving Barracks Camp Hollyday was established and named in honor of Rear Adm. Richard C. Hollyday. Camp Hollyday was designated a U.S. Naval Training Center in March 1944.

An aerial photograph of Gulfport in 1943 was taken shortly after the Bureau of Yards and Docks chose the base as the third Advance Base Depot because of its large municipally owned pier. The pier extended into the 30-foot-deep Gulfport channel, allowing large ships to move directly to dockside with only the aid of a pilot. At each depot, a receiving barracks was established to house and offer advanced training to transient Seabees.

A Seabee-recruiting truck containing a diorama illustrating the accomplishments of the Naval Construction Battalions traveled around the United States to recruit troops. Men already skilled in the various phases of construction work were enlisted into the Seabees. Recruits representing about 60 different trades were offered petty officer rates based on their civilian construction experience and their age.

A diorama recruitment trailer at a recruitment stop is pictured here during World War II. Navy recruiters traveled around the country passing out brochures and enlisting trained construction professionals into the Seabees. Per presidential order, direct voluntary enlistment into the Seabees ended on December 15, 1942, requiring all future procurement of military personnel through the Selective Service.

Seabees learn how to hoist a jeep for a trip across a creek during their advanced training at Advance Base Receiving Barracks Camp Hollyday in Gulfport, pictured here in 1944.

Seabees train in the different firing positions at the Advanced Base Depot Gulfport rifle range during World War II. All navy personnel who were physically fit for duty were required to learn to use a rifle and pistol effectively. After learning the mechanisms of the rifle and pistol, personnel were taught the various positions and how to aim properly in each position.

Seabees learn iron molding at Naval Training Center Gulfport, pictured here in March 1944. Casting is a manufacturing process by which a molten material is introduced into a mold, allowed to solidify within the mold, and then ejected or broken out to make a fabricated part. The skill of iron molding would be invaluable to Seabee steelworkers deploying to remote areas in the Pacific theater who were often short on supplies.

Seabee "surveymen," pictured here in 1944, train in the basic skills necessary to conduct location surveys for roads, airfields, pipelines, ditches, buildings, drainage structures, and waterfront construction. Surveymen, now called engineering aides, assisted construction engineers in developing final construction plans, conducting land surveys, and preparing maps, sketches, drawings, and blueprints. Seabee surveyors trained to operate and maintain various types of precision surveying and laboratory test instruments and equipment.

Members of the 103rd Naval Construction Battalion salvage a wrecked Flying Fortress in September 1943. Aircraft salvage was the process of rescuing an aircraft, its cargo, and sometimes the crew from peril. Salvaging of materials was especially necessary during World War II, when every major country imposed a system of rationing, price controls, and recycling of materials.

Seabees from the 80th Naval Construction Battalion aboard a 6-by-18-pontoon barge during the final boarding and stowing of gear are pictured here in 1943. After clearing the west pier at the Port of Gulfport, the Seabees headed for military training on Cat Island.

Members of the 80th Naval Construction Battalion aboard a pontoon barge prepare to land at Cat Island in 1943. As the battalion landed, the rifle squads moved to the forward section of the barge with the machine guns in the Jeeps behind the men. The Advance Base Receiving Base utilized Cat Island during World War II for training Seabee personnel in military combat tactics.

In June 1943, the 80th Naval Construction Battalion makes an amphibious landing at Cat Island as part of its military training in Gulfport. During World War II, the Cat Island War Dog Reception and Training Center, located on the island, trained dogs for military service in the U.S. Army Signal Corps.

This 1943 view shows numerous 1-by-12-pontoon strings moored at the small craft harbor near the Pontoon Assembly Yard at the Port of Gulfport. Various assemblies could be made from pontoon strings, including net tenders, warping tugs, causeways, rhino ferries, 75-ton floating crane barges, drydocks, finger piers, seaplane service piers and ramps, and aircraft landing fields.

The Pontoon Assembly Yard used by Advance Base Depot Gulfport to store materiel set for shipment to the Pacific theater is pictured here in 1943. The pontoon assemblies were the outstanding items of specialized gear developed by the Bureau of Yards and Docks during World War II. Assemblies, in the form of causeways and lighters, immeasurably aided troop and equipment landings in Sicily, Normandy, and on numerous Pacific islands.

Shown here in 1943 is a dining hall at Advance Base Depot Gulfport used during World War II. The commissary stewards, cooks, bakers, butchers, and jacks-of-the-dust aided the supply officer as was necessary for the purchase, stowing, issuing, and preparation of food for all hands. Jacks-of-the-dust are in charge of breaking out provisions for the food service operation.

Pictured here in 1943 is an Advance Base Depot Gulfport mess hall galley used to feed thousands of officers and enlisted men every day. A galley, originating from the galley ships of ancient times, is the compartment of a ship, submarine, train, or aircraft where food is cooked and prepared. A galley also refers to a land-based kitchen on a naval base.

The Pontoon Assembly Area at the Port of Gulfport was used for training Seabees in pontoon assembly during World War II. In order to create the large quantities of pontoons needed for barges, piers, and floating drydocks, Seabees trained at Gulfport to build and operate a pontoon assembly plant in the field.

Advance Base Depot Gulfport's open storage area housed equipment and materiel set for shipment to advance base depots around the world. Gulfport shipped 6,000 long tons of materiel by rail to New Orleans and other ports for trans-shipment in 1942. In 1943, there were 75,000 long tons shipped, and this doubled in 1944 to 154,000 long tons.

A warehouse located on Seventh Street at Advance Base Depot Gulfport was used to store materiel set for shipment to advance base depots in the Caribbean and Pacific theaters. Shipments of materiel and equipment from the advance base depots were based on directives either from the chief of naval operations (CNO) or on requisitions from advance bases in the field. Shipments made as a result of CNO directives usually were part of an initial move, such as the establishment of airfields and bases on an enemy-held island.

A storehouse at the Advance Base Depot, pictured here in 1944, stored materiel set for shipment to the Caribbean and Pacific theaters. A close liaison between the depot and the shipping authorities permitted a smooth flow of materiel from storage areas to dockside to shipboard.

Base public works installs new white shingles onto the Advance Base Depot Gulfport Administration Building No. 1. In March 1944, Advance Base Receiving Barracks was re-designated U.S. Naval Training Center and provided for several training schools, including basic engineering, diesel, radioman, quartermaster, and electrician ratings.

Seabees arrived by train on Pullman cars or day coaches at the Gulfport Union Station after traveling for days from Camp Endicott, Rhode Island; Camp Peary, Virginia; and other locations.

The base chapel, located in Building 169 between Bainbridge and Hull Avenues on Fifth Street, offered both Protestant and Catholic services and had two chaplains assigned to the center until 1945. The center's chapel remained closed from 1946 to 1952, when the Bureau of Yards and Docks took control of the base and renamed it Naval Construction Battalion Center Gulfport.

Seabees with the 100th Naval Construction Battalion march down the streets of Gulfport displaying several pieces of their heavy equipment in 1943. Seabee units participated in parades along the coast from New Orleans to Pascagoula during their brief stay at Advance Base Depot Gulfport.

In order to boost the sale of war bonds, the 100th Naval Construction Battalion was called upon to march in the streets of Gulfport, pictured here in 1943. Joining the army, marines, and WACS (Women's Army Corps), the Seabees made a good appearance, proving their versatility. The 100th Naval Construction Battalion served on Majuro Atoll, Marshall Islands; Angaur, Palau Islands; and Samar Island, Philippines.

Seabees and base civilian employees unload a boat at the east pier near the Pontoon Assembly Area in 1943. Most pontoons were shipped to the Pacific area in the form of flat plates and rolled shapes. Seabees in frontline Pontoon Assembly Detachments assembled the pieces into pontoons at mobile plants.

Seabees pose for a group photograph at Advance Base Receiving Barracks Camp Hollyday in December 1943. Group photographs of each platoon were taken at Camp Hollyday before the units shipped to Port Hueneme and the Pacific theater.

Barbers at the base barbershop provide haircuts and shaves to naval personnel. The barbershop is part of the ship's service, which operates many of the services aboard a ship or naval installation. During World War II, the ship's service managed to produce a nominal profit, which was used to promote the welfare of the crew through renting moving picture programs, purchasing athletic equipment, and making various other activities possible.

The Ship's Service Store at Advance Base Depot Gulfport, pictured here in 1943, operated with appropriated funds and was staffed by enlisted personnel. Bases and battalions can only create a Ship's Service Store when no other navy facility exists for selling merchandise. This counter sold only tobacco products for armed forces personnel and their families.

WAVE officer Lt. Symmie Gough of Austin, Texas, demonstrates the Colt .45-caliber automatic pistol at the Naval Training Center firing range here in 1945. The Colt .45 M1911, designed by John M. Browning, was the standard-issue sidearm for the U.S. Armed Forces from 1911 to 1985.

Naval Training Center WAVES stand at attention during captain's inspection here in February 1944. The WAVES (Women Accepted for Voluntary Emergency Service) began in August 1942, when Lt. Comdr. Mildred H. McAfee was sworn in as the first director of the WAVES and the first female commissioned officer in U.S. Navy history.

The Naval Training Center WAVES basketball team poses for a quick group photograph in 1944. The Bureau of Yards and Docks and the Naval Training Center used WAVES in clerical and administrative positions, releasing enlisted men and male officers for technical and advance base duties.

Seabees with the 25th Naval Construction Regiment stand at attention while stationed at Plymouth, England. On April 1, 1944, the 25th Naval Construction Regiment stood up for the principal function of planning, training for, and the execution of all projects in which the Seabees would be involved during the Normandy invasion.

A rhino ferry, operated by the 81st Naval Construction Battalion on D-Day, approaches the shoreline at Utah Beach, pictured here June 6, 1944. Rhinos towed behind LSTs (landing ship, tanks) arrived at 3:00 a.m. on D-Day off Utah Beach, where the sea was heavy and bombing and strafing by German planes added to the severity of operating conditions.

The 81st Naval Construction Battalion marries a rhino ferry to an LST during the invasion of Normandy on June 6, 1944. The battalion operated rhino ferries across the British Channel to the Normandy coast and the beach camps at Omaha Beach.

A rhino ferry is married to an LST during the Normandy invasion. To provide for the needs at Omaha and Utah Beaches, the Seabees at their British bases assembled 36 rhino ferries. In addition, they assembled 12 causeway tugs, 12 warping tugs, 2 rhino repair barges with a 5-ton crane, and 2 floating drydocks with enough capacity to dock an LST.

The 111th Naval Construction Battalion sets up the first Seabee camp on the hills above Normandy Beach, pictured here June 13, 1944. Seabees constructed and operated camps for naval personnel behind the invasion beaches. On D-Day plus seven (June 13), personnel were finally able to occupy pup tents erected in orderly rows along the lines of trenches in the bivouac area.

On March 22, 1945, Seabees with Construction Battalion Maintenance Unit 629 prepare landing crafts to transport Allied armored forces across the Rhine River at Oppenheim. To support Gen. George Patton's advancing army, the Seabees built pontoon ferries similar to the rhino ferries of D-Day fame and transported Patton's tanks across the river. In all, the Seabees operated more than 300 craft, shuttling thousands of troops into Germany.

African American Seabee divers with the 34th Naval Construction Battalion work on a marine railway at Gavutu, Solomon Islands. S1c. H. M. Douglas (left), Carpenter's Mate 2nd Class T. A. Blair (center), and Shipfitter 2nd Class I. Phillip are using improvised gas masks modified into diving gear to complete their task.

An interior view of an LTA (lighter-than-air) steel hangar built by the 80th Naval Construction Battalion at Carlsen Field, Trinidad, is pictured here on March 26, 1944. In the fall of 1943, the Seabees from the 80th Naval Construction Battalion deployed to Carlsen Field to construct naval lighter-than-air facilities in the Caribbean. To supplement the eight army-owned buildings taken over by the navy, the Seabees built a large, steel blimp hangar, a mooring circle, paved runways, a helium-purification plant, and other operational appurtenances.

Seabees on Guadalcanal wade through knee-high water to reach the mess hall. The Japanese airfield at Guadalcanal was close to completion when the American invasion forces struck. The principal task assigned to the 6th Naval Construction Battalion was to construct and keep the field operational during the months of fierce combat. The 6th Naval Construction Battalion followed the marines to Guadalcanal on September 1, 1942, and became the first Seabees to engage in the combination of building and fighting for which they have become famous.

The U.S. Marine Corps toast their Seabee road makers with this sign after the Seabees named one of their miraculous roads on Bougainville the "Marine Drive Hi-way." Shown shaking hands across the sign are navy Seabee Chief Boatswain's Mate Earl J. Cobb (left) and Marine Corps corporal Charles L. Marshall.

Comdr. Thomas Jones (center, dark clothing) and Lt. Comdr. John McAllister (center right, light clothing) with the 110th Naval Construction Battalion stand beneath the nose of a B-29 on Tinian admiring the Seabee insignia. At the height of the war, nose artists were in very high demand in the Army Air Force and were paid quite well for their services.

Seabees from Construction Battalion Maintenance Unit 591 created a handmade chest on Majuro Atoll, Marshall Islands, pictured here in 1944. The base at Majuro Atoll supported two marine dive-bomber squadrons, half of a patrol squadron, and a temporary staging area for one army fighter group. In addition, it provided Naval Air Transport Service requirements, fleet anchorage, medical facilities, and a loran transmitting station.

Officers and enlisted members of an Army Air Force B-29 and the 13th Naval Construction Battalion pose for a photograph on Tinian. The practice of personalizing fighting aircraft originated with Italian and German pilots. The first recorded piece of nose art was a sea monster painted on the nose of an Italian flying boat in 1913. While the nose art in World War I was

mainly embellished or extravagant squadron insignia, true nose art occurred during World War II, considered the golden age of nose art by many observers, with both Axis and Allied pilots taking part.

Members of the 34th Naval Construction Battalion host a luncheon for an African American USO troop on Guadalcanal, Solomon Islands, shown here in July 1944. A detachment of 260 men from the 34th Naval Construction Battalion, stationed at Tulagi, was ordered to Guadalcanal to take over the building of a 36,000-barrel tank farm at Koli Point.

Pictured here in September 1943, Seabees with the 22nd Naval Construction Battalion lay Marston matting on the east-west runway in Attu, Alaska. The Seabees arrived in Massacre Bay on May 21, 1943, to build housing for 7,000 men, materiel storage, seaplane landing areas, hangars and repair shops, drydocks, and maintenance shops.

Seabees unload cargo onto the black sandy beaches of Iwo Jima, pictured here in February 1945. Three Seabee units, assigned to the marines during the assault, operated as beach landing parties and later began working on the airstrips. The capture of Iwo Jima, halfway between Tokyo and Saipan, not only eliminated a base from which the Japanese could attack the Marianas, but it also provided a base for fighter planes escorting bombers to Japan.

Seabees with the 21st Naval Construction Battalion complete a freighter ship pier at Baten-Ko, Okinawa, pictured here on August 29, 1945. The freighter was the first ship to use the pier after the 21st Naval Construction Battalion finished construction. The pier head extension and second approach were installed as an addition to the original single-approach "T"-head pontoon pier used during the war.

Seabees with the 75th Naval Construction Battalion weld together fuel drums into a sewer line while stationed on Guam. Seabee ingenuity became legendary in the early days of World War II, when deployed units created hundreds of unique items out of surplus materiel and abandoned enemy equipment.

In the Philippines, Seabees attend Sunday chapel services at the base theater. Chaplains served with Seabee units from September 1942 until the end of World War II. Before hostilities ended, hundreds of navy chaplains served with the Seabees. At one time, about 135 chaplains had simultaneous duty with the Seabees.

Two

1946–1965
THE TRANSITION YEARS

Gate No. 1 (now the Pass Road Gate) is pictured in May 1954, shortly after the Bureau of Yards and Docks assumed administration of the base. In 1952, the base was renamed Naval Construction Battalion Center Gulfport, and two other components activities were simultaneously brought into being: the U.S. Advance Base Storage Depot and U.S. Naval Construction Equipment Depot. For 14 years, CBC Gulfport remained a Construction Battalion Center without a construction battalion.

Gate No. 4, pictured here in May 1954, at the west end of Naval Construction Battalion Center Gulfport, led into 223 acres of unimproved wooded area used for tactical training.

Pictured here January 14, 1954, the Bureau of Yards and Docks stored heavy equipment and materiel at Naval Construction Battalion Center Gulfport for immediate shipment to Seabee battalions around the world. Even though the population of the Seabee Center was drastically reduced after World War II, it still kept busy housing and restoring war reserve and strategic materiel. Storage of this equipment and materiel kept the center from closing after World War II.

An open storage area houses heavy equipment at Naval Construction Battalion Center Gulfport with warehouses in the background that store supplies for use in case of a national emergency. Pictured here May 4, 1954, the open storage facilities covered approximately 511,300 square yards of base land.

In 1952, the Naval Storehouse officially disestablished and became Construction Battalion Center (CBC) Gulfport, operated by the Bureau of Yards and Docks. By 1954, CBC Gulfport housed vehicles and heavy equipment in open storage areas. Open storage was used for materiel not easily affected by the weather or for materiel that was to be stored for a short time.

A recreation building used for athletics onboard Naval Construction Battalion Center Gulfport is pictured here in May 1954. Athletics remained an important factor in the development of good morale and were encouraged both ashore and afloat. Commanding officers were encouraged to increase morale and maintain physical fitness by forming as many athletic teams among various sports as was possible for their organization.

The Naval Construction Battalion Center Gulfport Officer's Club with messing facilities, shown here in May 1954, was used by commissioned officers as a place of recreation as well as a dining room. Usually a galley or scullery adjoins the officer's club and stewards provide service.

Pictured here in May 1954, the training aid projection building, located in the old Naval Training Center, was utilized by the base and reserve units. Countless films, produced as a training tool, supplemented classroom, technical training, and education.

The Naval Construction Battalion Center Gulfport display at Keesler Air Force Base on Armed Forces Day, May 15, 1954, promotes the Construction Equipment Depot and the Advance Base Supply Depot. The exhibition also included numerous pieces of heavy equipment stored at the center in case of a national emergency.

On May 15, 1954, the Seabees display a representation of heavy equipment at Keesler Air Force Base on Armed Forces Day. The 30 items of equipment chosen for display were picked from over 25,000 items valued at approximately $46 million, being held in stock reserve at Naval Construction Battalion Center Gulfport.

An Experimental Amphibious Transporter was used to conduct amphibious experiments at the base from December 1954 to April 1956. The Experimental Amphibious Transporter arrived at Gulfport by navy tow from Higgins, Inc., of New Orleans, Louisiana, on December 8, 1954.

Three

1966–1972
THE SEABEES COME HOME
TO GULFPORT

Seabees assigned to Naval Mobile Construction Battalion 128 lay the cornerstone at a children's hospital as part of a civic action project during their 1969 deployment to Vietnam. The Seabees are, from left to right, (first row) unidentified and Chief Rayburn Williams; (second row) Lt. Comdr. Donald Campbell, executive officer; Comdr. Joseph Gawarkiewicz III, commanding officer; CWO 2nd class Jack Masler; and unidentified.

An aerial view of downtown Gulfport highlights a strong mercantile center and historic home sites along the beach. In March 1966, when the center received its first construction battalion in 20 years, the city of Gulfport boasted a population of 38,000.

Pictured in 1969, the center's sign just inside the Pass Road Gate displayed the command logos of five battalions, one reserve battalion, and one regiment that stood up at Gulfport to support construction in Vietnam and globally. The sign was destroyed during Hurricane Camille in August 1969.

The Broad Avenue Gate and security station, used as the main entrance to the base, are pictured here in 1967, one year after the base reopened. In Gulfport, Naval Mobile Construction Battalions 62, 74, 121, 128, and 133 were commissioned in 1966 and 1967 to aid in meeting construction demands in Vietnam and Thailand.

Building No. 1, used by Battalion Center Gulfport's administration office and comptroller division, was located just inside the Pass Road Gate, pictured here in 1961. Although behind the scenes a good deal of the time, the administration office and comptroller department constituted one of the key parts of CBC Gulfport, providing the overall function of financing for the center.

The ceremony commissioning the 20th Naval Construction Regiment (NCR) at the Naval Construction Battalion Center Gulfport is shown here April 11, 1966. The 20th NCR provided training for up to seven construction battalions during the Vietnam-era buildup of the Naval Construction Force (NCF). The 20th Naval Construction Regiment served as the personnel receiving and processing activity for deployed Atlantic Fleet Naval Construction Force units.

Pictured here April 11, 1966, the administration building was utilized by the 20th Naval Construction Regiment on the day of commissioning. Wood-frame galleys, garages, and other utility structures of World War II vintage that had been idle for 20 years would become regimental training facilities at commissioning time.

Building 136 was utilized as an enlisted men's barracks for reserve construction battalions around 1970. World War II open-bay barracks were used to house temporary personnel, such as students, reserves, and cadets until additional barracks were built in the 1990s.

Pictured here July 25, 1966, Naval Mobile Construction Battalion (NMCB) 62 Charlie and Delta Companies receive instruction in railroad maintenance during home port training. From July to the end of September 1966, the men of NMCB 62 busied themselves in technical instruction related to their construction specialties and in military training in preparation for their first deployment to Vietnam.

A Seabee Team and its equipment are shown here. Seabee Teams were highly mobile, air transportable construction units that provided disaster relief and technical assistance to Allied nations. These teams could be tailored to fit any size task but normally consisted of a junior civil engineer corps officer, 11 construction men, and a hospital corpsman.

Seabees cross train in the skills of their colleagues and prepare for deployment with Seabee Technical Assistant Teams (STAT), later called Seabee Teams. These teams consisted of men specially selected from Naval Mobile Construction Battalions to provide direct technical assistance to friendly nations throughout the world. The teams proved exceptionally efficient in rural development programs and earned reverence as the "Navy's Peace Corps."

In 1967, a temporary CEC/Seabee Museum was established in Building 54 as part of the 25th birthday of the Seabees and the 100th anniversary of the Civil Engineer Corps. The museum exhibited items pertaining to Seabee and Civil Engineer Corps history, donated by both veterans and active duty personnel.

The Seabees created a mock Vietnamese village attack at the Naval Construction Battalion Center Gulfport Open House celebrating the 25th anniversary of the Seabees and the 100th anniversary of the Civil Engineer Corps, pictured here March 5, 1967.

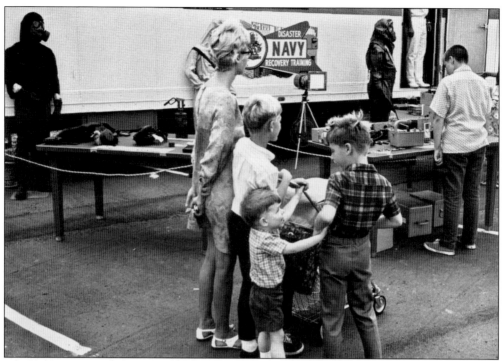

The Disaster Recovery Training department exhibits its equipment at the CBC Gulfport Open House on March 5, 1967. The department's main purpose was to provide naval personnel with training in nuclear, biological, and chemical recovery. The Disaster Recovery Training Program provided realistic field training to acquaint battalion personnel with some of the problems they must overcome in emergencies.

Seabee Team 6202 displays weapons captured from the Viet Cong and North Vietnamese in the Go Cong Province at the CBC Gulfport Open House on March 5, 1967. Seabee Team 6202 worked as teachers and supervisors for crews of local Vietnamese that labored to learn new skills, which would enable them to continue civic action projects long after the team left the province.

Naval Mobile Construction Battalion 74 utilizes map and compass training in a simulated battle at Keesler Air Force Base Training Area in July 1969. Seabees conducted tactical training at Keesler Air Force Base and Camp LeJeune Marine Base in North Carolina to help prepare for unexpected dangers that might appear in the coming months in Vietnam.

Mortar teams from Naval Mobile Construction Battalion 62 stand by as Maj. B. P. Cole of the 20th Naval Construction Regiment inspects a Seabee's service weapon during a 782 gear inspection in March 1969.

Naval Mobile Construction Battalion 62 personnel stand by for 782 Gear Inspection given by Commander Construction Battalions Atlantic (COMCBLANT) on March 5, 1969. During a 782 gear inspection, Seabees carefully lay out all personal combat gear issued while assigned to that unit, and each item receives scrutiny by staff officers, chiefs, and, upon occasion, honored guests.

Reserve Naval Mobile Construction Battalion (RNMCB) 12 personnel are educated by marine instructors in various kinds of rifle fire used to keep an enemy pinned down in a combat situation. During the first of four weeks at Camp LeJeune, RNMCB 12 received Marine Corps–type training with both the .45-caliber pistol and the M-14 rifle.

During field exercises at Keesler Air Force Base Training Area on February 18, 1969, Naval Mobile Construction Battalion 62 advances on a tactical march to the battalion defensive position. Tactical training included spending cold nights huddled in muddy holes waiting for aggressors to attack. However, the war games served a very real and important purpose in the essential training and conditioning needed before deployment to a combat zone.

The Naval Mobile Construction Battalion 62 field mess operates during field exercises on February 18, 1969. Mess specialists assigned to amphibious or naval mobile construction battalions prepared and provided food for the Seabees. Their duties include locating the site for the field kitchen, constructing it, setting up kitchen tents, arranging field equipment, and reloading field equipment.

Catholic Ash Wednesday services were held at the Naval Mobile Construction Battalion 62 campsite during field exercises in February 1969. Chaplain corps officers are responsible for the performance of all duties relating to religious activities. Chaplains conduct divine services, form voluntary religious classes, maintain liaison with the American Red Cross and Navy Relief, and investigate emergency leave, humanitarian shore duty, and hardship discharge.

Naval Mobile Construction Battalion 74 undergoes demolition training at Camp LeJeune in May 1969. During the last three weeks of a four-week course, Seabees trained in the Counter Guerilla Warfare School, the Demolition Course, or the Field Communications School at Camp LeJeune.

Surveyors from Naval Mobile Construction Battalion 62 lay out a road for the equipment operators during their home port training on August 3, 1966. Working on an extensive community service program, the Seabees undertook various projects in the local community, including a shoreline survey for a proposed lake boundary, finishing work on a teen center, and base improvements at the Seabee center.

Seabees from Naval Mobile Construction Battalion 74 train at Camp LeJeune on the 81-millimeter mortar, pictured here in May 1969. The M29A1 81-millimeter mortar is a smoothbore, muzzle-loaded, high angle-of-fire weapon and consists of a tube, bipod assembly, and base plate. The M29A1 medium mortar weighs about 98 pounds and can be broken down into several smaller loads for easier carrying.

Reserve Naval Mobile Construction Battalion 12 Seabees train in firing the M40 recoilless rifle during two-week active duty training at Camp LeJeune, North Carolina. The M40 recoilless rifle was a lightweight, portable, crew-served 105-millimeter weapon that primarily saw action during the Vietnam War. Mainly intended as an anti-tank weapon, the M40 was employed in an antipersonnel role with the use of the antipersonnel-tracer flechette round.

In April 1970, Seabees from Naval Mobile Construction Battalion 133 familiarize themselves with the M20 3.5-inch rocket launcher at Keesler Air Force Base rifle range. The M20A1 and M20A1B1 rocket launchers are two-piece smoothbore weapons with an open tube and are fired by an electrical firing mechanism, which contains a magneto providing the current. These launchers were designed to be fired from the shoulder in a standing, kneeling, sitting, or prone position.

Yeoman 2nd class Chuck Gwinn, training with his Seabee unit at Camp LeJeune, shoulders the M20 3.5-inch rocket launcher while wearing a gas mask. The M20 launches high explosive and smoke rockets against ground targets. These high-explosive antitank rockets are capable of penetrating heavy armor at angles of impact greater than 30 degrees.

Reserve Naval Mobile Construction Battalion 12 becomes familiar with Viet Cong clothing, gear, and equipment to aid recognition in combat situation. The Viet Cong were a guerilla force that battled the United States and South Vietnam during the Vietnam conflict.

Seabee members of Reserve Naval Mobile Construction Battalion 12 train how to identify a North Vietnamese army soldier in a replica Vietnamese village at Camp LeJeune, North Carolina. Members of the North Vietnamese army wore tan, long-sleeved, button-down wool tunics with pockets on either side of the shirt breast, tan wool pants, and a tan helmet with a brown strap.

Reserve Naval Mobile Construction Battalion 12 Seabees train in guerilla warfare at Camp LeJeune, North Carolina, in 1968. The Punji stick, or Punji stake, was a type of non-explosive booby trap used by the Viet Cong. Punji sticks were made out of wood or bamboo, placed upright in the ground, and covered by natural undergrowth, crops, or grass to camouflage their presence.

Seabees with Reserve Naval Mobile Construction Battalion 12 train in Viet Cong tactics, including a Punji stick–covered booby trap located in a tree at Camp LeJeune, North Carolina, pictured here in August 1968. The National Front for the Liberation of South Vietnam, also known as the Viet Cong, was an insurgent organization fighting the Republic of Vietnam during the Vietnam War.

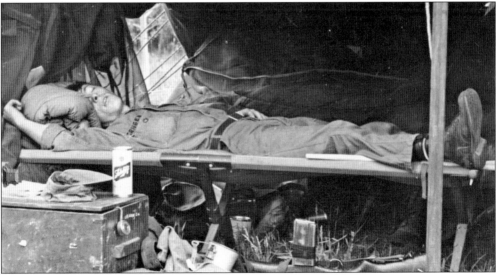

A Seabee rests during field exercises (FEX). Field exercises prepare the battalions and certify them as ready for possible future deployments and contingency operations outside the United States. Seabees train in tactical movement and combat operations, as well as force protection, chemical/biological/radiological (CBR) drills, reconnaissance, and patrols, and they learn to operate in a hostile field environment while sustaining construction and security operations through various exercise scenarios.

Seabees from the Direct Procurement Petty Officer (DPPO) Program carry a telephone pole during training at Naval Construction Battalion Center Gulfport in March 1969. Because of the growing construction requirements in Vietnam, the Naval Construction Training Unit trained skilled petty officers recruited to fill growing battalion ranks. The DPPO Program ended in February 1971 after training approximately 10,000 Seabees at Gulfport.

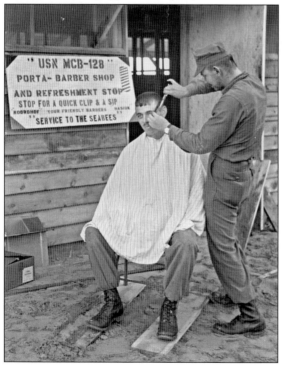

Ship's Serviceman Ira J. Noordhof, a member of Naval Mobile Construction Battalion 128, cuts Constructionman Builder Lloyd L. Aubel's hair at his Port-a-Barber shop in December 1967. Seabees working long hours in the field often found it impossible to get haircuts. To provide a solution, Seaman Noordhof went into the field three times a week to provide on-the-job haircuts.

Engineering Aide 3rd class Ed Sugg, assigned as soil tester for Naval Mobile Construction Battalion 62 Detail Buford, checks the moisture content of soil samples taken from the laterite fill of LTL-4. The density of the compacted fill is highly dependant upon proper moisture content. Laterite is a surface formation in hot and wet tropical areas; it is enriched in iron and aluminum and developed by intensive and long-lasting weathering of the underlying parent rock.

Seabees—from left to right, Engineering Aide 3rd class Ed Sugg, Equipment Operator 1st class Sam Bass, and Equipment Operator 2nd class Mike Morisoli—from Naval Mobile Construction Battalion 62 Detail Buford consume C-rations during a chow break in the back of a 5-ton cargo truck. Combat Individual, or C-rations, were combat rations issued by the United States during the Vietnam conflict and provided approximately 1,200 calories for each meal.

Seabees from Naval Mobile Construction Battalion 74 Detail Golf aid two members of a Seal Team that were ambushed in a boat one-half mile from the naval base at Nam Can, Vietnam, in early 1971. The Seabees heard the seriously wounded Seals scream for help and responded by giving first aid and carrying the men on stretchers to sickbay where a doctor was awaiting their arrival.

A convoy from Naval Mobile Construction Battalion (NMCB) 74 departs Camp Haskins for Phu Loc in U.S. Marine Corps vehicles to begin work on Route No. 1. NMCB 74 deployed to Camp Haskins in Da Nang, Republic of Vietnam, in June 1967. Largely in support of Military Assistance Command, Vietnam (MACV), NMCB 74 constructed an ammunition storage complex, galleys, living accommodations, roads, storm drains, and helicopter pads.

Engineering Aide 3rd class Jim Millar (kneeling) and the Naval Mobile Construction Battalion 62 detail survey nine miles in three days in 1968. A team of NMCB 62 surveyors spent four days in the field surveying a future road into Viet Cong–held territory. The team departed in helicopters from Phu Bai Airport and landed across the Dai Giang River in a rice paddy.

At a well site in Bao Trai, Vietnam, Construction Electrician 2nd class J. C. Clements (left), Construction Mechanic 1st class D. G. Jewell (center), and Utilitiesman 2nd class D. C. Schmidt work on the shaft of a pump assembly. Developing nations primarily use hand pumps as a means of bringing water to the surface from a borehole, rainwater tank, or well.

Fearless Ferris and Forces Fighting Raiders from Naval Mobile Construction Battalion 121's Alpha Company protect their road crew with a truck-mounted M60, pictured here in September 1967. NMCB 121 left Gulfport on July 28, 1967, for Phu Bai, Vietnam. While deployed, Alpha Company upgraded streets, built storage areas, reconstructed the Hue Causeway, and upgraded and widened a 29-mile stretch of QL-1.

Personnel from Naval Mobile Construction Battalion 128 Delta Company plow through a rice paddy as they set fence posts at Ammunition Supply Point No. 1, Da Nang, Vietnam. Delta Company provided perimeter security in the form of a 6-mile chain-link fence that was put up over terrain varying from swampy rice paddies to the sheer slope of a rocky hillside.

This is an aerial view of Pass Christian after Hurricane Camille swept through the streets in the late hours of August 17, 1969. Camille's winds were accompanied by 20-foot-plus tides that hit Pass Christian, nearly obliterating the community in its midnight swipe across the coast.

Navy Seabee Disaster Recovery Teams had the grim task of searching for victims in Pass Christian after the city reported almost total destruction in its lower areas in August 1969. Mississippi governor John Bell Williams directed Capt. James M. Hill, commanding officer of Naval Construction Battalion Center Gulfport, to take control of recovery efforts in Pass Christian and evacuate all community members who wished to leave.

In August 1969, Seabees help Pass Christian residents into trucks during a massive evacuation of the stricken community. The Seabees conducted a search-and-recovery sweep as far inland as necessary to aid Mississippi residents in evacuating the stricken areas.

Seabees cleared the streets of debris so that rescue teams could enter certain areas, as pictured here in August 1969. Live power lines required disconnection, broken water and gas mains were cut off, and a search was made for survivors. Navy corpsman and doctors followed Seabee recovery crews into the field to aid civilians injured in the storm.

Highway 90 is shown here in August 1969 the day after Hurricane Camille passed across the Gulf Coast. Camille destroyed or damaged 30,000 homes and hundreds of business, civic, and religious structures; severed communications; and knocked out water, power, and sewage services.

Wives of naval personnel prepare food for dependants and civilians in Warehouse 18, which served as a shelter during and after Hurricane Camille struck the Gulf Coast in August 1969. At Naval Construction Battalion Center Gulfport, 29 buildings were destroyed and more than 90 percent of the structures had major damage. Camille left only one barracks in livable condition to house the 2,500 Seabees who were on base the morning of August 18, 1969.

Naval Mobile Construction Battalion 62 passes in review, paying tribute to their fallen comrades Equipment Operator 3rd class Francis E. Camden Jr. and Constructionman Builder Murlin E. Boon, who died after Viet Cong mortars smashed into NMCB 62's camp at Phu Bai, Vietnam, in the pre-dawn hours of January 20, 1968.

Seven men from Naval Mobile Construction Battalion 133 receive Navy Achievement Medals in a ceremony before the men and color guard from Gulfport, March 5, 1968. Capt. R. C. Engram, commanding officer of CBC Gulfport, pins a medal on Chief John R. White.

Three

1973–1990
HUMANITARIAN ASSISTANCE AND
GLOBAL CONSTRUCTION

While on deployment in May 1990, personnel from Naval Mobile Construction Battalion 133 Detachment Samoa muster in lavalavas, the traditional garb of American Samoa. A 100-man detail deployed to American Samoa in February 1990 to provide humanitarian relief and emergency disaster assistance in the wake of Typhoon Ofa.

A view of downtown Gulfport highlights some of its many significant buildings. The Port of Gulfport is considered the most accessible port on the Gulf Coast. Gulfport is located just 12 miles from one of the world's major deepwater shipping lanes and is approximately midway between New Orleans and Mobile.

One of the main administration buildings used by home-ported Naval Mobile Construction Battalions is pictured here around the early 1970s. The old administration buildings are a stark comparison to the Stennis Complex now used as headquarters for Seabee battalions, navy and Marine Corps reserve, and base administration.

The base housing office staff, located in Building 54, seen here in the early 1970s, assisted all personnel in locating housing, both on and off the base. At this time, government-owned family housing at the center consisted of seven sets of married officers' quarters reserved for key personnel. For eligible enlisted personnel, there were 125 leased units dispersed within a 10-mile radius of the station.

After Hurricane Camille destroyed nearly all barracks on August 18, 1969, the center constructed numerous new bachelor enlisted quarters and bachelor officer quarters, pictured here in July 1971. Constructed of concrete block with brick veneer, the three new barracks built at CBC Gulfport could house up to 1,356 Seabees at any one time.

Naval Construction Battalion Center Gulfport Gas Station is seen on November 29, 1972, the day the center opened its new self-service pumps. Instructions for use were printed on a large sign near the pumps to educate patrons in basic operation of the new feature. By eliminating a service attendant, the price of regular gas dropped 2¢ to 30.9¢ per gallon.

On August 27, 1968, the center's commissary, located just inside Gate No. 1, opened for business in the old Naval Training Center's laundry building. Commissaries are an integral part of the military's pay and compensation package. The store provides a valuable financial benefit to its customers while enhancing the welfare of military personnel and their families.

The center's Clothing and Small Store (pictured May 9, 1972), originally located in Building 62, reopened after Hurricane Camille in Building 16, Section A. Small stores were established for keeping standard uniform items and certain other personal articles available for sale to naval personnel.

Seabee surveyors plot an area of the center's golf course, which opened in 1976. Trees, fabricated ponds, and sand traps surround the nine-hole course, which was laid out by golf professionals and constructed mostly with Seabee labor. Naval Mobile Construction Battalion 1 Alpha Company constructed nine greens plus a practice green after they returned from deployment to Okinawa. Bravo Company laid out the drainpipes and placed grade stakes, and Charlie Company built the last of the bridges across the canals.

On May 20, 1987, the Stethem Memorial Navy Lodge was dedicated in memory of Steelworker 2nd class (Diver) Robert Stethem. Assigned to navy Underwater Construction Team 1 in Norfolk, Virginia, Stethem was returning from assignment in Nea Makri, Greece, aboard TWA Flight 847 when it was hijacked by members of the Lebanese terrorist organization Hezbollah. On June 15, 1985, Stethem was singled out and killed by the hijackers when they learned he was a member of the U.S. military.

Until 1993, large hills of bauxite covered 56 acres of open storage on the base awaiting conversion to aluminum. In 1948, the munitions board selected the naval storehouse to be custodian of certain national stockpile material procured by the Emergency Procurement Service of the General Services Administration. In October 1948, the base began receiving imported bauxite ore shipped from the Netherlands East Indies (now Indonesia) to repay a war debt.

On August 9, 1974, Equipment Operator 3rd class Paul Graham washes a piece of Naval Mobile Construction Battalion 74 Alpha Company's equipment in preparation for Exercise Kennel Bear. Until 2005, Seabees regularly took part in this five-day field exercise hosted by the 30th Naval Construction Regiment.

A Seabee dependant is seen here on September 1, 1989, dropping aluminum cans into a truck as part of the Seabee center recycling program and the navy's Resource, Recovery, and Recycling Program. Recycling proceeds went toward environmental abatement projects and morale, welfare, and recreation programs. Navy-wide, the program netted $12 million during fiscal year 1990.

On July 1, 1975, Naval Mobile Construction Battalion 71 lowered its flag for the last time at the decommissioning ceremony held at CBC Gulfport. Re-commissioned in October 1966, NMCB 71 made two deployments to Vietnam followed by tours to Guantanamo Bay, Antarctica, and Bermuda. The battalion and its detachments completed construction in virtually every corner of the globe and gave rise to the motto "Seventy One, Second to None."

Pictured is the Seabee float at the Veterans of Foreign Wars (VFW) Bicentennial Parade on March 1, 1976, that carried the Seabee Queen. A 30-man flag unit from the Naval Oceanographic Unit 5, a color guard from NMCB 1 and 133, and two Amtracks joined the Seabee center float in the parade. On the reviewing stand (not identified in any order) were Lt. Gov. Evelyn Gandy, Brig. Gen. Charles Cooper, and many military, state, and local officials.

On March 1, 1976, Seabees march through downtown Gulfport at the VFW Bicentennial Parade. The Naval Construction Battalion Center, with over 500 marchers, provided the largest number of participants. A four-man bicentennial color guard led the center's part of the parade, followed by a navy recruiting truck displaying navy uniforms and the 2nd Marine Corps Division Band from Camp LeJeune.

The Auxiliary Security Forces first squad was called into action to fight terrorists in a realistic training exercise. The force, made up of supplemental personnel from non-deployable tenant commands, was designed to assist the center's security force in situations involving terrorists, civil disturbances, vehicle searches, and riot control.

Proud Seabees from Naval Mobile Construction Battalion 74 in full Military Oriented Protective Posture (MOPP) gear exit the confidence chamber after exposure to CS gas (tear gas) during a chemical, biological, and radiological (CBR) training refresher course. This exercise emphasizes the importance of properly wearing a gas mask or a protective mask, and an improper fit against the face is quickly revealed by the agent.

This October 8, 1974, view of the Construction Equipment Depot (CED) area was taken by personnel in a bucket truck. CED provides specified services to support the Naval Construction Battalion Center's mobilization contingency mission. In times of peace, CED assures readiness and reliability of Naval Facilities Engineering Command–managed materiel in storage.

In northern New York state, Naval Mobile Construction Battalion 62 trains in cold-weather tactics. The Seabees learned to use snowshoes, shoot M16s and M60s with cold hands and in stiff winds, and create a camp in a field covered with snow. Arctic construction projects included an emergency landing zone, a new road and parking lot, observation tower, and repairs to several bridges.

Gunnery Sgt. Jack Smart instructs Seabees in proper cleaning techniques for the M1911A1 .45-caliber service pistol and service rifle. General disassembly and assembly is necessary for normal care and cleaning. Each battalion has a marine assigned as a military advisor to oversee military training and ensure all necessary information concerning military matters is conveyed to the chain of command.

Seabees jump out of their vehicle during their annual field exercise (FEX) to Camp Shelby. FEX is a valuable training experience for every battalion. It exercises Seabees' ability to mobilize quickly and provides a measurement of how well they coordinate a massive relocation into a contingency environment. FEX includes comprehensive technical, military, and operational training in the field, communications system instruction, and light weapons training.

Seabees practice setting up the 60-millimeter mortar at CBC Gulfport. After four days and one evening of class work and simulated training exercises, the men had the opportunity to become familiar with the weapon aboard the Seabee Center. After two days, the Seabees traveled to Camp Shelby for a three-week training session using live ammunition.

Seabees train on the 81-millimeter M1 mortar, pictured here in 1973. The M1 mortar provides air assault, airborne, ranger, and light infantry rifle companies with an effective, efficient, and flexible weapon. The 136-pound M1 was designed to be man-portable when broken down into three 45-pound components: the tube, the bipod mount, and the round base plate.

Seabees cross a three-strand bridge at the beginning of the endurance course at the Jungle Warfare Training Center at Camp Gonslaves, Okinawa. The endurance course is a 3.4-mile obstacle course, which is part of a weeklong jungle skills training class Seabees often complete while stationed in Okinawa. The Jungle Warfare Training Center teaches jungle warfare and skills, combat tracking, medical trauma, and survival evasion resistance and escape.

The former Advance Base Training Camp was dedicated as Camp Hill on July 12, 1972, to honor the memory of Capt. James M. Hill, former center commanding officer. Camp Hill is situated on a 575-acre plot of ground under lease to the navy in DeSoto National Forest, 15 miles north of Gulfport.

After withstanding a rigorous 782 gear, weapons, and personnel inspection by marines from the 20th Naval Construction Regiment Military Training Division, the Seabees were ready to deploy aboard aircraft from the Air Force's 815th Airlift Squadron at Keesler Air Force Base. The standard 782 gear is divided into three categories: fighting gear, bivouac, and protective equipment.

The battalion gunnery sergeant inspects an M-16 to ensure there is not a round in the chamber. A marine gunnery sergeant infantry specialist is assigned to each battalion to provide advice and expertise in the areas of military training, small arms, defense tactics, and liaison with tactical units.

Naval Mobile Construction Battalion 133 off-loads equipment from a Lockheed C-5 Galaxy. The C-5 Galaxy, part of the U.S. Air Force Air Mobility Command, is a military transport aircraft designed to provide strategic heavy airlift over intercontinental distances. It is the largest American military transport designed to carry oversized cargo and one of the largest military aircraft in the world.

Naval Mobile Construction Battalion 74 Detail Pohang departs Okinawa for Gulfport on February 6, 1990. Detail Pohang constructed troop housing at the Marine Expeditionary Camp in Pohang, Korea. The detail deployed in late December, and its team consisted of Lt. Steve Fischer, Chief Steelworker Bill Frantz, and 24 enlisted personnel of various ratings.

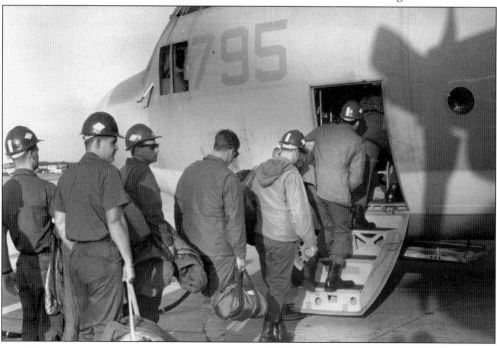

An aerial view of U.S. Naval Support Facility Diego Garcia in the Indian Ocean shows the extensive construction projects completed by the Seabees. One of the largest peacetime projects ever undertaken by the Seabees was the complete development, construction, and operation of Naval Support Facility Diego Garcia. The facility provided a defense communications network and furnished improved communications support in the Indian Ocean for ships and aircraft of both the U.S. and British governments.

Seabees with Charlie Company maneuver a 1.5-cubic-yard concrete bucket over building base forms by means of a crane. Organized to perform most of the battalion's vertical building projects, Charlie Company forms and finishes concrete projects and performs light frame and heavy timber construction and masonry work.

Seabees use ropes to lift and guide the K-span panels for proper placement. The K-span building system is assembled from panels preformed by an on-site automatic building machine, which turns coils of steel into structural strength arched panels. The 40-by-130-foot structure is seamed together to form an economical and watertight steel building. The Seabees have been using the K-span building since its introduction during Operation Desert Storm.

As part of Deep Freeze 73, Seabees deployed to Antarctica to construct the 6-story-high geodesic dome at South Pole Station. The Antarctic deployment of NMCB 71 resulted in the completion of the Siple Station project, erection of the geodesic dome and other structures at the South Pole, and completion of numerous projects in the McMurdo Station area. The geodesic dome covered and protected most of the buildings at South Pole Station.

Showing off their newly installed street-post sign, Naval Mobile Construction Battalion 74 continues the tradition of naming the streets after the battalion at Camp Shields, Okinawa, pictured here February 1990. The original idea for naming the streets came from Command Master Chief Richard Wade, who decided, while stationed at Camp Shields with NMCB 1, to name the unidentified roads at Camp Shields.

Family and friends turn out to welcome home Naval Mobile Construction Battalion 74 from deployment. The battalion re-commissioned December 6, 1966, at the Naval Construction Battalion Center in Gulfport and adopted the motto "Does More." Since 1966, NMCB 74 has seen four deployments to Vietnam, deployed around the world to conduct humanitarian and civic action construction, earned nine "Best of Type" awards, four Peltier Awards, and the Presidential Unit Citation.

On May 7, 1991, hundreds of family members and friends crowded the airport for the arrival of the first flight carrying members of Naval Mobile Construction Battalion 74 as they returned to Gulfport.

Five

1990–2001
GULF WAR TO 9/11

Naval Mobile Construction Battalion 133 Detachment Juliet-Echo displays the battalion's logo during morning quarters on April 1, 1996. The commanding officer briefed the battalion on their upcoming deployment to Bosnia, where they took part in Operation Joint Endeavor. As part of its peacekeeping mission, the Seabees constructed tent cities for American forces, built wooden walkways, and installed security perimeter fences.

Pictured here in March 1992, the administration building for CBC Gulfport and the 20th Naval Construction Regiment, part of the Stennis Seabee Complex, is the main focus of attention upon entering the main gate of the center.

The Stennis Seabee Complex, dedicated to Sen. John C. Stennis, encompasses two battalion headquarters buildings and the base command headquarters, pictured here in March 1992. Senator Stennis represented Mississippi for more than 40 years and served on the Senate Appropriations Committee, the Defense Appropriations Sub-Committee, and the Senate Armed Services Committee.

The center's 50th anniversary celebration took to local streets in the Gulfport Mardi Gras parade on March 3, 1992. Two Seabees get into the spirit of Mardi Gras by throwing beads from the Seabee float.

Seabees present arms during a weapons inspection at CBC Gulfport. The commanding officers often hold periodic command inspections to determine deficiencies and cleanliness. Quarters or formations are for the purpose of ceremony, inspection, muster, instruction, or passing orders.

Reserve Seabees sharpen their skills with the M16 rifle at the Camp Shelby firing range. Weapon training is part of the two-week active duty training period. The Seabee qualification course requires qualifying in all four standard shooting positions with a machine gun: prone, standing, kneeling, and sitting.

The Military Training Department at the 20th Naval Construction Regiment gives Seabees hands-on training using a service rifle and pistol. Training ensures the maximum number of personnel qualify as sharpshooters. The Military Training Department's mission is to provide combat readiness training to the battalions on base and all reserve units east of the Mississippi River.

Judges watch as several Naval Reserve Seabees reassemble their M60 machine guns during the Reserve Naval Construction Force Readiness Rodeo competition at Construction Battalion Center Gulfport on May 1, 1991.

Hospital Corpsman 2nd class Joseph McGee instructs battalion crews on the MK19 at the anti-tank weapons range during their recent field exercise, pictured here November 4, 1996. The MK19 is a belt-fed automatic 40-millimeter grenade launcher, or grenade machine gun, that entered U.S. military service during the cold war and remains in service today. The MK19 is a man-portable, crew-served weapon that can fire from a tripod-mounted position or from a vehicle mount.

Steelworker Constructionman Jeffrey Gill checks the alignment of K-span sections before permanently connecting them with the seaming machine. The final shape and strength of the materials used cancels the need for columns, beams, or any other type of interior support. All of the panel-to-panel connections are joined using an electric automatic seaming machine. Because of this, there are no nuts, bolts, or any other type of fastener to slow down construction or create leaks.

Pictured here in May 1993, Seabees with Naval Mobile Construction Battalion 1 spread hot asphalt on a road in Rota, Spain. Seabees deployed to U.S. Naval Station Rota have supported the base with public works and construction projects, as well as local coastal communities with civic action projects.

Seabees from Reserve Naval Mobile Construction Battalion 20 board a C-130 for field exercises. More than 600 reservists arrived on the Gulf Coast to begin two weeks of readiness duty at Camp Shelby in De Soto National Forest. Training included erecting huts, grading roads, laying sewers, and stringing electrical wires. The reservists also qualified with the M-16 rifle and practiced crew-served machine guns, mortars, and grenade launchers.

Seabees assigned to Naval Mobile Construction Battalion 74 take every precaution to ensure the equipment fits into the aircraft properly while loading a C-130 aircraft, pictured here in February 1992.

Seabee equipment, newly painted to United Nations standards, is loaded onboard the vessel SS *Mosa* in September 1997. Naval Mobile Construction Battalion 1 deployed a 179-person detail from Camp Mitchell, Rota, Spain, to Bosnia-Herzegovina to provide construction engineer support for Operation Joint Guard.

Ninety Seabees from Naval Mobile Construction Battalion 74, originally deployed to Puerto Rico, set up tents for Haitian refugees at U.S. Naval Station, Guantanamo Bay, Cuba. During Operation Sea Signal, navy and Marine Corps personnel assumed the mission of feeding, housing, clothing, and caring for more than 50,000 Haitian and Cuban migrants seeking asylum in the United States.

Seabees from Naval Mobile Construction Battalion 74 train local citizens in a humanitarian aid project. Humanitarian and civic assistance projects have historically represented one means for the United States operationally and medically to train its troops as well as aid new and underdeveloped countries. Civic action projects included repairing roads, drilling wells, building airfields, creating tent cities for refugees, and bringing aid to disaster victims around the world.

Naval Mobile Construction Battalion 1 clears debris at the Miami Heights Elementary School in Florida in August 1992. Seabees arrived three days after Hurricane Andrew hit and encountered a scene many described as a war zone. Within hours of arrival, Seabees were sent out to clear roads, clean up schools, and assist where needed.

The 20th Naval Construction Regiment pulled out an entire allowance of heavy equipment for Seabee battalions—including bulldozers, front-end loaders, and wreckers—for shipment to the Persian Gulf as part of Operation Desert Storm/Desert Shield in September 1990. To speed up materiel pack up, each piece of equipment carried a bar code to track it from storage to final destination.

Heavy equipment at the Gulfport pier awaits the Kuwaiti ship *Danah* for shipment to the Persian Gulf in September 1990. During this packout for Desert Storm, Construction Equipment Department (CED) personnel pulled all equipment out of storage, de-preserved, repaired, or upgraded, and repainted each to a desert scheme prior to shipment.

Seabees assigned to Naval Mobile Construction Battalion 74 pose with their marine liaison Gunnery Sgt. Michael LaBlue in Saudi Arabia. Just before the ground attack started, a group of 21 Seabees participated in deception tactics with a Marine Corps unit. Under the cover of darkness, Task Force Troy moved to within a mile of the southern border of Kuwait and set up mock tanks and artillery. The mission deceived the enemy regarding the true location of coalition ground forces.

Naval Mobile Construction Battalion 74 lay matting in Bahrain during Desert Shield, pictured here in January 1991. While detailed to Bahrain, the unit successfully completed several critical jobs, including installing utilities, constructing roads, and supporting structures for an army field hospital. The unit also improved the facilities to support the influx of military people.

A female Seabee helps another brush the wind-blown knots out of her hair during the Gulf War. The 1991 Gulf War proved to be the pivotal time for women in the U.S. armed forces. Over 40,000 women served in almost every role the armed forces had to offer. However, while many came under fire, they were not officially permitted to participate in ground engagements. Despite this, 16 women were killed in action and 2 were taken prisoner.

Builder 2nd class Carl Weinberg stands watch during Operation Desert Storm. On August 7, 1990, Seabees were notified that they would be included in Operation Desert Storm and immediately began packing up construction equipment for shipment to Saudi Arabia. The training undergone at home port proved invaluable, as the Seabees successfully supported the First Marine Expeditionary Force's mission in the Persian Gulf.

Coalition power team prepares a mobile substation for use in Saudi Arabia during the Gulf War. Mobile substations are self-contained units, which include not only the transformer, but also all high- and low-voltage switchgear.

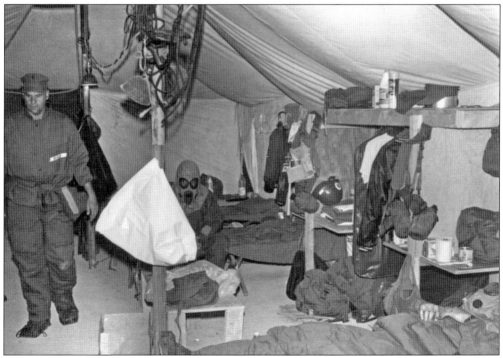

Seabees relax in their berthing tents after a long day of construction work in Saudi Arabia. The Seabees worked seven days a week, in two 12-hour shifts, with only Thanksgiving and Christmas off. On a typical day, they awoke at 5:30 a.m., ate breakfast, worked 12 hours, showered, ate dinner, wrote a letter, and were in bed by 8:30 p.m.

Naval Mobile Construction Battalion 74 enjoys Christmas dinner in the desert during the Gulf War on December 25, 1990. The 16 mess specialists and mess cooks served approximately 1,100 Seabees and marines three meals per day from a small makeshift galley in the middle of the Saudi Arabian desert.

Seabees with Naval Mobile Construction Battalion 133 operate a group of 621G Caterpillar scrapers in Sikh, Iraq, as part of Operation Provide Comfort. A wheel tractor-scraper is a piece of heavy equipment used for earthmoving. The rear part has a vertically moveable hopper with a sharp horizontal front edge, which, when lowered, cuts into the soil like a cheese-cutter and fills the hopper.

A Seabee operates the backhoe attached to the rear of a loader as members of NMCB 133's air detachment repair the runway at an Iraqi airfield on April 1, 1991. The Seabees repaired the airfield to allow relief supplies flown in for nearby Kurdish refugees as part of Operation Provide Comfort. The battalion convoyed personnel and equipment 400 miles from Iskenderun, Turkey, to Zakho, Iraq—the longest inland convoy operation conducted by a Seabee battalion since the Vietnam War.

Constructionman Electrician Maurice Troutman, from Naval Mobile Construction Battalion 133, performs electrical work during deployment to Rota, Spain. The battalion re-deployed shortly after arrival in Spain to Sikh, Iraq, for Operation Provide Comfort. Seabees deployed to aid Kurdish refugees abandoning their mountain strongholds for Displaced Citizen Communities (DCC) supervised by the United Nations.

A Seabee from Naval Mobile Construction Battalion 133 uses a rough terrain forklift to move a pallet of supplies after his unit's arrival in Iraq for Operation Provide Comfort, a multinational effort to aid Kurdish refugees in southern Turkey and northern Iraq. The Seabees were dispatched to Iraq from their base at U.S. Naval Station, Rota, Spain.

Pictured in January 1993, marines and Seabees work together to build landing and staging areas for CH-53 helicopters and a taxiway for C-130 aircraft for the multinational relief effort Operation Restore Hope in Somalia. Naval Mobile Construction Battalion 1 constructed the facility in three weeks using 600,000 square feet of AM-2, a type of aluminum matting.

A member of Naval Mobile Construction Battalion 1 works to secure a basketball hoop erected in a Somalian school playground in February 1993. The Seabees installed a basketball hoop and improved the playground of a local school as part of a civic action project during the multinational relief effort Operation Restore Hope.

In February 1993, two members of Naval Mobile Construction Battalion 1 put their initials in the concrete footing they just placed during Operation Restore Hope. The Seabees went ashore to provide construction support to the U.S. contingent and provided vertical construction support to U.S. and coalition forces establishing base camps at each of the humanitarian relief sites.

As part of Operation Joint Endeavor, Seabees of Naval Mobile Construction Battalion 133 construct wooden walkways at the base camp north of the town of Brka, Bosnia. NMCB 133 deployed a 198-person air detachment to Hungary, Croatia, and Bosnia-Herzegovina and deployed an 18-person team to Sarajevo to support the Implementation Force commander's headquarters (IFOR).

Seabees from Naval Mobile Construction Battalion 133 hold quarters at Camp S. A. Harmon, pictured here December 26, 1995. On Christmas day 1995, NMCB 133 Seabees arrived in Croatia as part of Operation Joint Endeavor, the peacekeeping effort in the former Yugoslavia.

These 51 Seabees from Naval Mobile Construction Battalion 133, shown on February 9, 1996, built Camp McGovern, the new home for the army's 3rd Battalion, 5th Cavalry, just north of the town of Brka, Bosnia. NMCB 133 deployed to bare land, constructed their camp and security perimeter, and then built the new camp in 11 days as part of the operation.

Three Seabees attached to Naval Mobile Construction Battalion 133 keep warm beside a kerosene space heater inside one of the newly constructed tents at the camp. In case of an emergency, Seabees rotated watches inside a tent to safeguard loaded M16A1 rifles for Seabees working nearby.

Although departures and homecomings are common occurrences for Seabees and their families, the return of personnel from Saudi Arabia after the Gulf War attracted more attention than usual. In May 1991, twenty-one Seabees from Naval Mobile Construction Battalion 74 returned to Gulfport with Naval Mobile Construction Battalion 24 after a full tour in Saudi Arabia.

In May 1991, friends and family members welcome Seabees home from Naval Mobile Construction Battalion 24's advanced party after their deployment to the Middle East during the Gulf War. The battalion was recalled to active duty in early December 1990 and deployed to Saudi Arabia. While deployed, the Seabees prepared sites and installed utilities for Fleet Hospital 15, constructed port access roads, and converted passenger buses into ambulance litter carriers.

Six

2001–2007

COUNTERTERRORISM AND
RECONSTRUCTION

On May 15, 2004, Seabees attend a memorial service in Fallujah, Iraq, honoring seven fallen Seabees with Naval Mobile Construction Battalion 14. On April 30, 2004, two Seabees died when an improvised explosive device (IED, or land mine) exploded as they drove by on patrol, and five Seabees died on May 2, 2004, during a mortar attack in Fallujah, Al-Anbar Province, Iraq. (U.S. Navy photograph by Photographer's Mate 2nd class Eric Powell.)

Seabees from Naval Mobile Construction Battalion 74 Delta Company march towards the Seabee Lake onboard the Naval Construction Battalion Center on May 22, 2006. Delta Company completed a hydration exercise that consisted of marching three miles in full gear to help prepare them for the heat and strenuous duties anticipated during an upcoming deployment. (U.S. Navy photograph by Photographer's Mate 2nd class Gregory N. Juday.)

Jiffy Team members from Naval Mobile Construction Battalion 1 stand by to clean the simulated chemical-infected area during a CBR exercise on February 7, 2006. NMCB 1 was preparing for their upcoming field exercise, Operation Steel Shanks, which sharpened the battalion's combat and contingency construction capabilities. (U.S. Navy photograph by Photographer's Mate 3rd class Ja'lon A. Rhinehart.)

Hospital Corpsman 2nd class Danny L. Hawkins, assigned to Naval Mobile Construction Battalion 74, slides across a river using a pulley system on August 17, 2005, as part of an obstacle course at the Jungle Warfare Training Center (JWTC), located in the Northern Training Area on the island of Okinawa, Japan. (U.S. Navy photograph by Photographer's Mate 2nd class Eric S. Powell.)

On August 15, 2005, Construction Electrician 3rd class Brian A. Neilsen rappels down a 65-foot cliff during jungle warfare training in the Northern Training Area on the island of Okinawa. Petty Officer Neilsen was assigned to Naval Mobile Construction Battalion 74. (U.S. Navy photograph by Photographer's Mate 2nd class Eric S. Powell.)

Steelworker Constructionman Joseph Perkovich (left) and Builder Constructionman Elliot Galloway, both assigned to the Naval Mobile Construction Battalion 7 air detachment, carry debris left over from the destruction the tsunami caused to a primary school in Ahangama, Sri Lanka, on January 16, 2005. The Seabees were assigned to clear the debris as part of the humanitarian relief efforts of Operation Unified Assistance, the humanitarian relief effort to aid the tsunami victims. (U.S. Navy photograph by Photographer's Mate Greg Bingaman.)

Seabees assigned to Naval Mobile Construction Battalion 1 help clear trees downed by Hurricane Katrina in the Gulfport area on September 4, 2005. Seabees supported Hurricane Katrina recovery in the Gulf Coast as part of Joint Task Force Katrina in support of Federal Emergency Management Agency (FEMA). (U.S. Navy photograph by Photographer's Mate 3rd class Ja'lon A. Rhinehart.)

Construction Battalion Center Gulfport Seabees discuss the actions needed to remove debris left by Hurricane Katrina from U.S. Highway 90, pictured here August 30, 2005. (U.S. Navy photograph by Photographer's Mate 3rd class Ja'lon A. Rhinehart.)

U.S. Navy Seabee vehicles from CBC Gulfport prepare to remove debris left by Hurricane Katrina on August 30, 2005. (U.S. Navy photograph by Photographer's Mate 3rd class Ja'lon A. Rhinehart.)

The front view of the CBC commissary shows the extent of damage from Hurricane Katrina, shown here on September 4, 2005. The commissary was severely damaged when Hurricane Katrina, a Category 4 hurricane, came ashore on August 29, 2005. (U.S. Navy photograph by Photographer's Mate Airman Paul Williams.)

Construction Mechanic 1st class Jesus Reyna, assigned to Naval Mobile Construction Battalion 7, takes time to eat lunch with his family in an equipment warehouse located at the Naval Construction Battalion Center on September 3, 2005. Military dependants and civilian workers from the center relocated to the equipment warehouses as a temporary shelter following Hurricane Katrina. (U.S. Navy photograph by Photographer's Mate 2nd class Michael Sandberg.)

Ensign Michael Dobling (center) feeds four dolphins with Institute for Marine Mammal Studies trainers on September 17, 2005. The four dolphins were placed in a temporary saltwater pool facility at CBC Gulfport after spending two and a half weeks in the Gulf of Mexico. The dolphins were swept out to sea from the safety of their aquarium in Gulfport by a wave reported to be 40 feet high during Hurricane Katrina. (U.S. Navy photograph by Journalist 3rd class Chris Gethings.)

Seabees assigned to Naval Construction Training Center (NCTC) Gulfport lay new pavement for a parking lot at St. Thomas Elementary on December 20, 2005. NCTC helped the school recover from the loss of their facilities because of Hurricane Katrina. (U.S. Navy photograph by Yeoman 1st class Brandon J. Ruppert.)

Pictured here on January 7, 2006, Seabees from the Naval Mobile Construction Battalion 1's First Class Association move shingles to the roof of a house damaged by Hurricane Katrina. NMCB 1 First Class Association chose community service projects as its key focus while in home port. (U.S. Navy photograph by Photographer's Mate 3rd class Ja'lon A. Rhinehart.)

Personnel assigned to Naval Mobile Construction Battalion 74 remove debris from ankle-deep water off the coast of southern Mississippi during the Great American Cleanup event on March 2, 2006. NMCB 74 joined volunteers from the surrounding community and across the nation in a continuing effort to clean up and rebuild the Gulf Coast region following Hurricane Katrina in 2005. (U.S. Navy photograph by Journalist 1st Class Rob Wesselman.)

Seabees assigned to Naval Mobile Construction Battalion 74 clear debris to allow for expansion of Dewan Tent City at Muzaffarabad, Pakistan, on November 5, 2005. Dewan Tent City was a temporary shelter for displaced earthquake survivors. (U.S. Air Force photograph by Airman 1st class Barry Loo.)

U.S. Navy Seabees assigned to NMCB 74 clear debris at the Ministry of Education Building in Muzaffarabad, Pakistan, on December 1, 2005. The U.S. government participated in a multinational humanitarian assistance and support effort led by the Pakistani government to bring aid to victims of the devastating earthquake that struck the region on October 8, 2005. (U.S. Navy photograph by Photographer's Mate 1st class Eric S. Powell.)

Seabees assigned to Naval Mobile Construction Battalion 74 place concertina wire around their compound in Muzaffarabad, Pakistan, on October 27, 2005. (U.S. Navy photograph by Photographer's Mate 1st class Eric S. Powell.)

Naval Mobile Construction Battalion 7 and Naval Mobile Construction Battalion 21 Seabees use a screed on freshly placed concrete that will be a generator pad at Camp Arifjan, Kuwait, on May 31, 2006. (U.S. Navy photograph by Photographer's Mate 3rd class Paul D. Williams.)

Pictured here January 16, 2002, military personnel walk past latrines constructed by the Seabees at a forward operating base located at Kandahar International Airport in Kandahar, Afghanistan, during Operation Enduring Freedom. (U.S. Marine Corps photograph by Capt. Charles G. Grow.)

Seabees of Naval Mobile Construction Battalion 74 pre-position steel medium-girder bridging for future river-crossing operations on February 10, 2003. (U.S. Navy photograph by Photographer's Mate 1st class Brien Aho.)

Seabees attached to Naval Mobile Construction Battalion 74 place concrete for a C-130 aircraft staging area on January 3, 2003. The C-130 staging area was the largest single-battalion concrete project undertaken since World War II and was used to support future operations within the Central Command Area of Responsibility. (U.S. Navy photograph by Photographer's Mate 1st class Brien Aho.)

Utilitiesman 3rd class Eduardo Rivera-Gonzalez, assigned to Naval Mobile Construction Battalion 74, fires a 7.62 mm M240B machine gun down range during weapons qualifications in Iraq on April 10, 2004. (U.S. Navy photograph by Photographer's Mate 2nd class Eric Powell.)

Engineering Aide 2nd class Oliver Taylor, assigned to Naval Mobile Construction Battalion 74, helps secure a sector of a highway after sighting an improvised explosive device (IED) on April 14, 2004. Taylor was part of NMCB 74's Seabee Engineer Reconnaissance Team, which deployed to central Iraq in support of Operation Iraqi Freedom. (U.S. Navy photograph by Photographer's Mate 2nd class Eric Powell.)

Iraqi children wave to an American convoy that passes on a local road on January 24, 2005. Seabees assigned to Naval Mobile Construction Battalion 7 worked in partnership with residents of a small Bedouin village on the outskirts of Najaf, Iraq, to build a school and implement improvements to the village water, electricity, and sanitation facilities. (U.S. Navy photograph by Photographer's Mate 3rd class Todd Frantom.)

Seabees hand out cookies to local Iraqi children in Najaf, Iraq, on January 24, 2005. (U.S. Navy photograph by Photographer's Mate 3rd class Todd Frantom.)

The Naval Mobile Construction Battalion 74, Tactical Movement Team escorts a construction crew convoy through Fallujah, Iraq, on April 6, 2004. (U.S. Navy photograph by Photographer's Mate 2nd class Eric Powell.)

Utilitiesman 3rd class Mark Crubaugh (right), attached to Naval Mobile Construction Battalion 1, works alongside marines from the 9th Engineer Support Battalion, 3rd Marine Logistics Group, III Marine Expeditionary Force, to assemble a medium-girder bridge during a training exercise at Camp Hansen, Japan, on November 14, 2006. Joint military training evolutions help to promote relationships between the marines and Seabees in contingency environments. (U.S. Navy photograph by Mass Communication Specialist 3rd class Ja'lon A. Rhinehart.)

In Isabela, Philippines, Builder 2nd class Michael Schneider (far left), attached to Naval Mobile Construction Battalion 7, places concrete with his Philippines navy Seabee counterpart, Fireman 1st class Builder Elmer Ang (far right), pictured here on June 4, 2007. They are two of over 50 Seabees from both countries working on an engineering civic action project for the students and teachers of Kuampurnah Elementary School. (U.S. Navy photograph by Mass Communication Specialist 1st class Dave Gordon.)

Rear Adm. Robert L. Phillips, commander, First Naval Construction Division, presents Steelworker 3rd class Christopher Moran with a Purple Heart at Naval Construction Battalion Center, Gulfport, on May 8, 2006. The Seabees were injured in Al-Asad, Iraq, while deployed with Naval Mobile Construction Battalion 133 in support of Operation Iraqi Freedom. (U.S. Navy photograph by Photographer's Mate 2nd class Gregory N. Juday.)

Families and members of Naval Mobile Construction Battalion 133 reunite at the Air National Guard Training Center in Gulfport on April 11, 2006. Naval Mobile Construction Battalion 133 and other detachments were forward deployed to central command areas of responsibility in support of the global war on terrorism. (U.S. Navy photograph by Photographer's Mate 1st class Shane T. McCoy.)

Family and friends watch as Seabees from Naval Mobile Construction Battalion 133 return to Gulfport on April 11, 2006, following a six-month deployment to Iraq. (U.S. Navy photograph by Photographer's Mate 1st class Shane T. McCoy.)

www.arcadiapublishing.com

Discover books about the town where you grew up, the cities where your friends and families live, the town where your parents met, or even that retirement spot you've been dreaming about. Our Web site provides history lovers with exclusive deals, advanced notification about new titles, e-mail alerts of author events, and much more.

MADE IN THE USA

Arcadia Publishing, the leading local history publisher in the United States, is committed to making history accessible and meaningful through publishing books that celebrate and preserve the heritage of America's people and places. Consistent with our mission to preserve history on a local level, this book was printed in South Carolina on American-made paper and manufactured entirely in the United States.

This book carries the accredited Forest Stewardship Council (FSC) label and is printed on 100 percent FSC-certified paper. Products carrying the FSC label are independently certified to assure consumers that they come from forests that are managed to meet the social, economic, and ecological needs of present and future generations.

FSC
Mixed Sources
Product group from well-managed forests and other controlled sources

Cert no. SW-COC-001530
www.fsc.org
© 1996 Forest Stewardship Council

Find *Your* Place in History.